ALFRED DICKIE'S UTILITY BILL

INNOVATION AND
TECHNOLOGY

ALFRED

DICKIE'S

UTILITY BILL

Discussion Guide

Includes Story, Q&A and Facilitator's Tips

ENERGYPHILE SESSION №3

ENΞRGYPHILE

Published by Energyphile Media Inc.
energyphile.org

ISBN 978-1-9991113-8-0 (paperback)
ISBN 978-1-9991113-9-7 (ebook)

Produced by Page Two
pagetwo.com

Edited by Lori Burwash
Cover and interior design by Taysia Louie
Original design concept by Christina Sweetman

Contents

PREAMBLE

CAN A UTILITY BILL from 1915 describe the issues that assail our energy circumstance today? "Alfred Dickie's Utility Bill" is one of the first short stories I wrote. It's also one of my favorites. By the time I was done writing, I was surprised at the richness of meaning embedded in the simple scrap of paper I'd picked up for a dollar at an antique market.

My research into the Halifax Electric Tramway Company's invoice gave me insight into how people want their energy, what prompts someone to switch to a new system (from wood to gas, in Alfred's case) and how technology reduces the cost of energy over time.

Above all, Alfred's bill helped me understand an important turning point in our energy history: the point at which we lost touch with where and how we get our energy. In turn, I realized why people today are so distrustful of companies that supply our fuels.

Alfred used to know what it took to heat his home. After 1915 he didn't. Now think about where your energy comes from. Do you know what generates the electricity behind your wall plug?

Where the gasoline in your car originates? The natural gas in your furnace? What would it take for you, like Alfred, to switch to a new energy source, such as installing solar panels on your roof?

Alfred Dickie's utility bill symbolizes much more than a $1 indebtedness to a utility company — it marks a foundational story about how new sources of energy alter societal behavior and, yes, it aptly describes the issues that assail our energy circumstance today.

A rubber stamp with a 1915 date often has a fascinating story absorbed in its faded ink.

ALFRED DICKIE'S
- UTILITY BILL -

I'D LIKE YOU to meet Alfred Dickie. Why? Because you'll be surprised by what he can teach us about our energy choices.

I came across Alfred's utility bill at an antique fair, the type of place where small treasures are begging to tell big stories. Leafing through tattered magazines, faded sheet music and other bric-a-brac, I found a bluish piece of paper from the Halifax Electric Tramway Company. My interest was piqued when I saw the date: November 30, 1915.

A rubber stamp with a 1915 date often has a fascinating story absorbed in its faded ink. I've long been captivated by the period between 1880 and 1920. It was the glory era of energy transitions: we traded loyal horses for emotionless cars, noisy steam engines for smelly diesel motors and glowing lanterns for electric light bulbs. And don't forget the Wright brothers' first flight at Kitty Hawk in 1903. By 1915, this momentum had pushed the early adoption of electrical grids and petroleum products. With each trade-in, our lifestyles and economy changed immeasurably.

DISCOUNT will be allowed only if bill is PAID on or before TENTH of Month

ALL ACCOUNTS ARE PAYABLE AT THE OFFICE OF THE COMPANY.

Office Hours, 9 a. m. to 5 p. m.
Saturday, 9 a. m. to 1 p. m.

5213

ALFRED DICKIE,
95 SOUTH PARK ST.

HALIFAX, N. S.
Nov. 30th, 1915

To THE HALIFAX ELECTRIC TRAMWAY CO., Ltd., Dr.

FOR GAS CONSUMED DURING PAST MONTH:

TO NOV. 27 METER INDEX 647 .00
FROM OCT. 28 METER INDEX 637 .00

MINIMUM RATE $

10 HUNDRED CUBIC FEET AT $1.25 PER M. $ 125

20 PER CENT. DISCOUNT IF PAID ON OR BEFORE 10th DEC. 25

NET $ 100

STOVE ACC.

RECEIVED PAYMENT FOR THE COMPANY

ARREARS

DEC. 1915

TOTAL $

NOT RESPONSIBLE FOR LEAKAGE
Bill Must be PAID within 20 DAYS or Service will be Discontinued
CUSTOMERS MUST NOTIFY THE COMPANY OF ANY CHANGE IN ADDRESS

OFFICE: 13 LOWER WATER STREET

PRESENT THIS BILL WHEN MAKING PAYMENT

Alfred's 1915 utility bill gives us a glimpse into the era when coal gas was a new fuel piped into homes for heating, cooking and lighting — a luxury afforded only by the wealthy.

Turning over Alfred's bill, I nodded — drawings of a stove and a water heater confirmed I had something of value. Flipping it back, I noted Alfred's particulars: he lived at 95 South Park Street in Halifax. Having visited Nova Scotia's capital many times, I imagined his house was in the late-Victorian style: a foundation of brick or concrete shouldering a couple of stories, with a covered front porch.

Handwritten meter readings recorded that Alfred had consumed "10 hundred cubic feet" of heating fuel in November 1915. For this, he was obliged to pay $1.25 "within 20 DAYS or Service will be Discontinued," according to the rather terse Halifax Electric Tramway Company.

Thankfully, he paid within three days. His speedy December 2 payment, acknowledged by a PAID rubber stamp, earned him a reward of 25¢. Nice, I thought: pay early and earn a 20% discount. These days the mentality is reversed: pay late and earn a penalty. I wonder, When in our energy history did the attitude shift from carrot to stick?

A buck went a long way a hundred years ago, so I didn't know if Alfred's bill was cheap or expensive for that brisk Atlantic month of November. Regardless, his payment of $1 seemed fitting because, a century later, my antique dealer was asking the same amount — for a piece of paper most people would have thrown out decades ago. Not me, of course.

As I flipped the bill over a few times, I pondered. Who was Alfred Dickie and what could we learn from the decisions he made a century ago? What was the Halifax Electric Tramway Company and why was it so keen to promote gas? Relative to today, was $1.25 a little or a lot to pay for gas?

I looked up and said, "I'll take it."

The soppy-faced antique dealer accepted my dollar without hesitation. That dollar bought me a treasure trove for my curiosity.

The Lumber King

I had occasion to go to Halifax in 2014. Alfred's bill in hand, I found some answers. South Park Street is still there, green and leafy as it doubtless was in 1915. However, Number 95 is no longer, most likely demolished during construction of the Victoria General Hospital.

Alfred Dickie was a prominent Canadian businessman, nicknamed the Lumber King of Nova Scotia. After attending Dalhousie College in Halifax, he returned home to Stewiacke and went into the lumber business. He wasn't the first to see value in Nova Scotia's forests, but historical accounts suggest he was one of the best at harvesting the renewable resource.

Alfred's operations grew rapidly. He bought up three large forest properties (18,000 acres here, 65,000 there) and a number of sawmills and launched several pulp and lumber companies. He even owned a few steamships. Soon, the Lumber King became the leading exporter of Nova Scotia forestry products.

But his fortunes didn't last. Like so many capitalists in a volatile commodity business, Alfred overextended himself. Bad luck — fires and harsh weather, aggravated by a lack of insurance — and a tough business cycle contributed to his downfall. The banks took over and Alfred lost all but one share of his company's equity.

Despite his misfortunes, he remained a director of his holdings ... and was evidently paid well to do so. In the early 1900s, his salary was recorded as $833 per month. Now, $800 doesn't sound like much but, adjusted for 100 years of inflation, Alfred's annual income today would have been almost $250,000 — not bad for a bankrupt CEO. He was hardly suffering financially and, like any determined entrepreneur, he worked to recover his wealth and station. In 1912, Alfred went into semi-retirement and moved into that Victorian house on South Park Street.

Knowing that Alfred was wealthy through boom and bust gave me valuable context for his $1.25 gas bill. I discovered that while he was earning his $833 monthly paycheck, those Nova Scotia workers who sported overalls and lunch buckets earned 25¢ an hour — less than $45 a month. While a dollar for piped-in gas was loose change from Alfred's pocket, it would have been a considerable chunk out of the wallet of any hardworking fisherman, bricklayer or carpenter.

But the numbers weren't adding up.

I returned to the amount of gas — "10 hundred cubic feet," or 1,000, as it would be written today. A modern North American household with a furnace, water heater and stove consumes between 8 and 10 times that much gas. Alfred was either miserly with his energy use (particularly so in the cold month of November) or his house had only one or two small gas-fired appliances. The latter made more sense.

According to the bill, the Halifax Electric Tramway Company offered "Gas Appliances of all kinds." Because Alfred's usage was so low, I suspected he'd installed either a stove or a water heater. Neither appliance consumed much compared to a gas furnace heating an entire home. In fact, gas furnaces were not common in 1915 because most homes were still heated by burning wood. So Alfred's low-volume utility bill represented an early stage of transition to the gas era in Halifax.

The other side of Alfred's bill advertised the superlative virtues of coal gas — and promoted some appliances while it was at it.

Alfred's low-volume utility bill represented an early stage of transition to the gas era in Halifax.

Now We're Cooking with Gas!

The history of the Halifax Electric Tramway Company is intriguing. This contemporary energy utility came into being through a series of corporate consolidations. Three entrepreneurs — Benjamin Franklin Pearson, Frederick Stark Pearson (no relation) and Henry Melville Whitney — foresaw opportunity in providing citizens with alternatives to chopping wood, riding horses and burning candles. What an incredible time to have business acumen. I imagined the trio's excitement as they realized they could make a fortune converting Halifax to the modernity of trams, electrical appliances and heat from gasified coal.

Thanks to their foresight, Pearson, Pearson and Whitney did indeed get rich, and no wonder. Think about what they were offering: instead of going outside in the cold, finding your ax, chopping a tree trunk and hauling in the wood (which would take several minutes to light and still longer to give off heat), you could simply turn the gas valve and flick a match. Today we take the luxury of instant, hassle-free energy for granted. Yet, in 1915, it was life-altering. A gas flame wasn't just faster and more convenient; it was clean and touted as safe, without the nuisance of a wood fire's smoke and leftover cinders.

With that lure, the well-to-do citizens of Halifax had good reason to hop on the city's electric tram and travel down to "Gas Lane," where they could visit the small showroom of the Halifax Electric Tramway Company. I could just imagine a young sales buck in that showroom, coaxing families like the Dickies into installing a gas line and buying a shiny new cast-iron stove. The confident salesman probably added, "May I interest you in a water heater too, sir?" Seeing the benefits, the Dickie household must have been easily convinced to switch from wood to gas, and presumably from lanterns to light bulbs too.

In fact, the marketing claims on the back of Alfred's bill were compacted into four superlatives: "Cheapest, Safest, Cleanest, Quickest." Not "Cheap, Safe, Clean, Quick," not even "Cheaper, Safer, Cleaner, Quicker." No, the Halifax Electric Tramway Company's gas appliances apparently topped all other energy alternatives.

It goes without saying that consumers still want energy to be the cheapest, safest, cleanest and quickest. We want affordable energy that doesn't pick our wallets, safe energy that doesn't harm our person, clean energy that doesn't degrade our planet and quick energy that doesn't test our patience for warmth, light or transport at the flick of a switch or turn of a key.

"Cheapest" has become even cheaper for people like you and me. Adjusting for inflation, we pay about one-third for our gas of what Alfred paid for his. The delivery of our gas is safer and more reliable too, especially after most utilities migrated from coal gas to natural gas in the 1930s.

My attention turned to society's present-day situation: if, a century ago, we had already transitioned to the "cheapest, safest, cleanest, quickest" energy source, where does that put us today?

I think we can agree that consumer attitudes have hardened considerably since Alfred's bill. In the face of unsettling issues like climate change, environmental protection and not-in-my-backyard resistance, the list of superlative adjectives we require in our energy sources is growing. Now we're looking for such claims as "lowest carbon," "fewest pipelines" and "greatest social conscience."

That thought left me unsettled. I put Alfred's bill on my desk and stared out the window. Power lines, street lamps and cars were unavoidably in my view. Surely, the list of required superlatives for our energy can't grow ad infinitum, or even beyond the length of what can be advertised on a simple energy bill. Yet why do we seem to think the list can grow?

Trading Blisters for Convenience

Picking up that blue paper again, I realized something big had happened when the people of Halifax transitioned to direct-to-the-home energy services. Utilities like the Halifax Electric Tramway Company coaxed citizens into surrendering a primal survival chore: going out and getting their own supplies.

When someone like Alfred Dickie chopped his own wood, brought it into his house and lit his kindling, he was in control of his energy, from the wood pile to the stove. He knew if he was running out of logs, if his fuel was not cut to size or if it was too wet to burn. By necessity, people of his era had a vital connection to their energy sources, especially wood. They knew how to use it wisely and were acutely aware of changes in its availability. Any incompetence or oversight could lead to cold meals at best, a family freezing to death in the middle of winter at worst.

Maybe I'd stared at the bill too long, but I had a revelatory thought. The clerk who stamped Alfred's bill with a dull rubber thud marked a symbolic moment, when consumers and suppliers lost a shared sense of responsibility for society's energy needs.

Over the last century, switches, valves and thermostats have made our lives convenient. But those knobs and dials have also pushed energy sources "upstream," out of our homes, and out of our minds, to distant horizons where pipelines and wires terminate. To be sure, separating source from usage has led to comfort and prosperity. But when we ceased getting blisters from a sweaty ax handle, we unwittingly became detached from the complexities of getting heat and light at the flick of a switch.

From the suppliers' perspective, I note the lesson of Alfred's bill in the difficulties today's energy companies face with securing approvals to build things like pipelines, power plants or even wind farms. Sure, they are all regulated into supplying the safest and cleanest according to government specs. And each competes against the other to be the cheapest and quickest. But whatever they do, companies discover that being the best is still never good enough for the most demanding fringes of the populace.

If, a century ago, we had already transitioned to the "cheapest, safest, cleanest, quickest" energy source, **where does that put us today?**

What will narrow this functional imbalance between suppliers and demanders?

Okay, I admit that the gap to achieving "best" is vague. That's intentional, because I can't predict what will satisfy everyone's needs, short of a perpetual motion machine that delivers free energy, forever, with no pollution.

I do know that a realistic "best" offering will have to come from energy suppliers and consumers rebuilding a common mindset. We don't live in a world where the driver of an SUV has to work on a drilling rig to bring a barrel of oil out of the ground. But we can live in a world where oil and pipeline companies interact better with consumers, giving the person behind the steering wheel a stronger grip on the complexities of getting gasoline to the pump. Such interaction would benefit any company, government or investment group trying to introduce or build a new energy system, renewable or otherwise.

Come Together

At the start of this story, I suggested you might be surprised by what we can learn from Alfred Dickie's utility bill. Alfred chose to transition from his wood-based system to the latest source, gas. By signing up with the Halifax Electric Tramway Company, he outsourced his barehanded duties and gained convenience.

Today, we take that convenience for granted, and have all but relinquished our ability to provide our own energy. While we don't need to start chopping our own wood again, I believe that to make truly responsible choices in the future, we must all — consumers and suppliers — reacquaint ourselves with the challenges of demanding and providing superlative-laden energy.

From my career in energy, I know it's not realistic for consumers to take a crash course in supply-side economics. I also know that most suppliers have lost touch with their end-use customers. Nevertheless, where there are business problems, there are winners waiting with solutions.

I made another business trip to Halifax in July 2015, almost a hundred years to the day after Alfred's utility bill was issued. I had some time before my flight home, so I rented a car and drove an hour north to the Pine Grove Cemetery in Stewiacke, where Alfred was laid to rest in 1929. A large personality in life, he accordingly had the tallest gravestone in the cemetery. The golden-tinged granite seemed to insist that he'd died a wealthy man.

Ironic, I thought: someone who'd made so much money selling wood — the Lumber King himself — was a lead adopter of a fuel rival. That salesman on Gas Lane must have been pretty good at gaining Alfred's trust. That, and having the lure of cheapest, safest, cleanest and quickest to back up his pitch.

It just goes to show that those who understand energy superlatives, and how to build trust around them, will be the leaders of the future.

FACING In death, as in life, Alfred looms large.

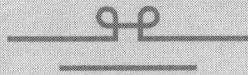

QUESTIONS
AND ANSWERS

Introduction

Sometimes a small curiosity can lead to a discussion about big issues. Alfred Dickie's utility bill, a hundred-year-old slip of paper, is such an item. The questions in this section aim to get you talking about:

- where our energy comes from
- the factors that induce us, or not, to buy new energy appliances
- the role that wealth and income play in the adoption of new energy systems
- how to think about trade-offs among the benefits we expect from our energy supplies
- the economic and financial considerations of transitioning to new energy systems

There's a lot to talk about. For fun, bring one of your own utility bills to the discussion and refer to it as you discuss the questions. When you're done, ask yourself if much is different today, a hundred years after Alfred Dickie paid his bill.

Questions

1 Alfred Dickie's story speaks to how the rise of distant energy suppliers (like gas utilities) separated people from their energy sources. As a result, society has become dissociated from understanding the connections between sources of supply and consumption — and, by extension, from understanding

the sources of environmental problems like greenhouse gas emissions.

A Look around the room: notice the lights, heating/cooling vents, electrical sockets, computers, maybe even a stove or fridge. Where does the energy that powers these amenities come from? Is it from one source or many?

B Thinking about your energy usage, discuss where greenhouse gas emissions, if any, are generated.

C When it comes to tackling climate change by reducing emissions, where are the bulk of the policy forces targeted? Why?

D What current energy system brings consumers closer to their source?

E What are some challenges in educating the public on energy supply?

2 Unlike today's impetus to reduce emissions by switching to renewables, the transition from wood to gas was not considered urgent a hundred years ago. Upstart utilities like the Halifax Electric Tramway Company were motivated by the opportunity to commercialize new technology. As such, Alfred Dickie didn't receive a government subsidy to make a move to gas heating. Nor was the wood he burned subject to a "wood tax."

A Why do consumers today receive subsidies to install solar panels on their homes or buy electric vehicles (EVs)?

B Why didn't Alfred get a subsidy?

C Do you think a wood tax would have influenced Alfred to switch to gas faster?

3 Affluent people are more inclined to buy new, exciting products, regardless of cost, probably because they can easily afford to acquire such high-value products. (This helps explain why "the Lumber King" was an early adopter of expensive gas heating and a new appliance.) The same can be said of wealthy companies and countries. At a personal, corporate and global level, the affluent demographic's early adoption helps to usher in energy transitions.

A What's a current example of wealthy people being the first to adopt new green energy–consuming products even if they're more expensive?

B What about organizations? First, think about examples around the world, then discuss what your organization is doing, if anything, to adopt new energy technologies.

C At the global level, what countries have been leaders? To what benefit?

D Discuss the role of income inequality in facilitating energy transitions.

4 One observation in "Alfred Dickie's Utility Bill" is that, over the course of a hundred years, the real, inflation-adjusted price of gas delivered to the consumer declined by 75% (from $25 per cubic foot to less than $7, depending on location). Think about the long-term price of energy as you answer these questions.

A Discuss the reasons for the deflationary trend.

B Do you expect the real price of energy to continue its decline over the next hundred years? Why or why not?

c Discuss the economic and political factors that can limit for-ever declining energy prices.

D What are the negative consequences of continually declining consumer energy prices?

5 We have high expectations of our energy supplies, so much so that we tend not to think about how the lights go on, or even if the lights *will* go on. We simply assume they will. Just as we assume our monthly utility bill won't bankrupt us. The Halifax Electric Tramway Company marketed its new energy product as "Cheapest, Safest, Cleanest, Quickest," without any discrimination as to the relative importance of each benefit. Yet what if you had to choose between reliability and safety of your energy supply? Or cost and cleanliness?

A What attributes are most important to customers buying appliances today?

B Do we take for granted the amenities that energy brings to our lives?

c Today's definition of "clean" is far different from what it meant a hundred years ago. In Alfred Dickie's day, it meant your house didn't fill with smoke and you didn't have to scoop ashes out of heating appliances. Now, "clean" is dominantly about reducing greenhouse gas emissions. Are you willing to trade off "cheapest, safest and quickest" for being a zero-carbon emitter, or "cleanest" in today's terms?

6 The success or failure of a new product depends on many factors. Investors will assess a product's viability based on one set of parameters, while the market may determine success based on other criteria.

A Imagine you're an investor in 1915. You have a wood stove at home but notice that a few friends are buying gas stoves and appliances. You also note the Halifax Electric Tramway Company is installing more pipes into some neighborhoods and wonder if you should invest in this exciting new utility venture. What considerations would you take into account before investing in the gas business?

B What was it about gas stoves that made them so successful over wood?

C What are some new appliances or products that didn't succeed on the market?

D What have you learned about new product adoption?

7 Macroeconomic factors — circumstances in the broad economy and capital markets — are also influential in the adoptability (or not) of new energy systems. Think about these factors for the following questions.

A As a reward for paying his bill early, Alfred Dickie was given a fairly hefty discount: 25¢ on $1.25. Why do you think the Halifax Electric Tramway Company offered such a discount?

B What are the external influences that companies offering new energy solutions today must contend with?

Answers

1 A Where does the energy that powers the amenities in your room come from? Is it from one source or many?

You'll probably point out the lights first. That's good, because understanding where electricity comes from is a great place to start when thinking about energy sources. Depending on where you live, the mix of primary sources used for generating power is diverse, and the electrons that flow through the sockets are generated from sources that include wind, solar, hydro, nuclear, natural gas and probably coal too.

In Canada, 82% of electrical power originates from renewable hydroelectric power — an extraordinarily high percentage relative to most other countries. The remaining 18% of Canada's electricity comes from a combination of nuclear (14.8%), coal (8.7%) and natural gas (8.6%).

Yet Canada is a big country with diverse geographical and jurisdictional variations. If you live in Alberta, 89.7% of the province's electrical power comes from coal and natural gas. On the other hand, in British Columbia and Quebec, over 90% is hydro. In Alfred Dickie's home province of Nova Scotia, power generation comes from a mix of coal (55.9), natural gas (16.7%), wind (11.8%), hydroelectric (8.8%) and other sources (6.8%).

For general global reference, the table on page 22 provides a select country-by-country breakdown of the energy mix used in electrical power generation in 2018.

Electricity is only one form of energy we use. When it comes to heating, many buildings in cold climates use heating oil (a petroleum product), propane or natural gas in their furnaces.

ELECTRICITY GENERATION BY FUEL, 2018*

	OIL	NATURAL GAS	COAL	NUCLEAR ENERGY	HYDRO-ELECTRIC	RENEW-ABLES	OTHER**	TOTAL TERAWATT-HOURS
Canada	0.5%	9.0%	9.1%	15.3%	59.2%	6.9%	0.1%	654.4
US	0.6%	35.4%	27.9%	19.0%	6.5%	10.3%	0.3%	4,460.8
Brazil	2.0%	8.0%	3.7%	2.7%	65.9%	17.8%	0.0%	588.0
Germany	0.8%	12.8%	35.3%	11.7%	2.6%	32.2%	4.5%	648.7
United Kingdom	0.5%	39.4%	5.0%	19.5%	1.6%	31.6%	2.3%	333.9
Russian Federation	1.0%	46.9%	16.0%	18.4%	17.1%	0.1%	0.4%	1,110.8
Saudi Arabia	39.2%	60.7%	0.0%	0.0%	0.0%	0.0%	0.0%	383.8
China	0.2%	3.1%	66.5%	4.1%	16.9%	8.9%	0.2%	7,111.8
India	0.6%	4.8%	75.3%	2.5%	8.9%	7.8%	0.0%	1,561.1
Japan	5.7%	36.8%	33.0%	4.7%	7.7%	10.7%	1.5%	1,051.6
TOTAL WORLD	3.0%	23.2%	38.0%	10.2%	15.8%	9.3%	0.6%	26,614.8
OECD	1.7%	28.4%	25.6%	17.5%	12.6%	13.0%	1.1%	11,233.6
Non-OECD	4.0%	19.4%	46.9%	4.7%	18.0%	6.6%	0.2%	15,381.2
European Union	1.6%	18.9%	20.0%	25.2%	10.5%	21.5%	2.3%	3,282.2

* Based on gross output.
** Includes sources not specified elsewhere (e.g., pumped hydro, non-renewable waste) and statistical discrepancies.
Source: *BP Statistical Review of Energy 2019* (see Sources Cited on page 54).

Do you know where your heating energy comes from? Quite likely it's from a distant oil or natural gas field, or a blend of fuel coming from many such fields half a world away. As in Alfred Dickie's circumstance, your fuel is probably refined at a plant and distributed by a utility via pipeline or truck (the latter in the case of heating oil).

"Alfred Dickie's Utility Bill" points out that people who live on the utility grid, with pipes and wires coming into their homes, are mostly unaware of where their energy comes from. That's understandable. No longer do we make candles from animal tallow, nor chop wood to heat our homes or cook our food. Our sources of energy come from afar through complex infrastructure, delivered just in time, on demand.

The complexity of the electrical power delivered on the grid is amplified because the sources of power generation are many. You could be charging your mobile phone from a nuclear power plant, a gas-fired natural gas generator and a wind turbine at the same time. It's not easy to comprehend, and the opacity of how it all happens tends to make us take for granted these complex, behind-the-scenes energy systems.

1 B **Thinking about your energy usage, discuss where greenhouse gas emissions, if any, are generated.**

If your electricity is, even in part, produced at a central source like a power plant fueled by coal or natural gas (which it probably is), combustion emissions are generated at that facility. Carbon dioxide is the dominant greenhouse gas emission blown out of big chimney stacks.

On the other hand, if the electricity mix that comes into your home includes power generated from solar panels, wind turbines, nuclear plants or hydro dams, there are no operating emissions.

If you cook on a natural gas–fueled stove or heat your home with an oil or gas furnace, the bulk of emissions are created in that location, where the fuel is burned. Similarly, for gasoline- or diesel-powered vehicles, the engine under the hood burns the fuel, which means those emissions are blown out the tailpipe.

The bulk of greenhouse gas emissions from fossil fuels are produced at the point of combustion. For example, for automobiles, 80% of emissions are produced in the engine, with only about 20% in the processes that extract, refine and distribute the fuel to the gas station.

☺ FACILITATOR'S TIP

Knowing where emissions are produced is important to designing effective policies for reducing them. Ask the group to discuss the difference between emissions generated at centralized locations like power plants and those produced in decentralized points, such as homes and cars. Which source is more difficult for policy to mitigate?

---

1 C **When it comes to tackling climate change by reducing emissions, where are the bulk of the policy forces targeted? Why?**

For the most part, policy continues to try to force a transition to renewable energy at centralized sources of emissions, like coal-fired power plants. This makes sense because effecting a change at a big generating facility reduces emissions on a large scale.

Where policy has a greater challenge is in energy-consuming segments of the economy, like transportation, where the bulk of emissions are generated in the internal combustion engines of "decentralized" vehicles. Legislating change in this domain is much more difficult because millions of people would have to trade in their vehicles, say for lower-emission hybrids or zero-emission all-electric models. Getting millions of people to change their consumption behavior is clearly more challenging than swapping out one big power plant.

As a means of tackling these energy-consuming segments, policy initiatives like carbon taxes have been front and center. Such levies on personal wallets aim to raise the price of gasoline so that drivers are encouraged to switch to electric cars. In homes and commercial buildings, carbon taxes on natural gas are intended to get people to make more energy-efficient choices when it comes to furnaces, appliances, windows and insulation and generally be more mindful about their day-to-day usage. Carbon taxes at an industrial source, like a coal-fired power plant, are designed to force utilities to switch to lower-carbon or zero-carbon electricity generation like natural gas or renewable sources, respectively.

1 D **What current energy system brings consumers closer to their source?**

Solar panels are a great example of a system that brings people closer to the energy source that powers their home. Meters and controls allow homeowners to monitor electrical power generation and consumption. With in-home power storage (like batteries), the homeowner is also apprised of how much is in reserve. Having energy sources closer to home makes users more conscious of conservation. When you know where your energy comes from, you tend to have a greater respect for energy use and sustainability.

Solar panels on a home have a couple of other benefits. First, economics: solar panels have come down in price sufficiently to compete with electrical power purchased from the grid (the wires coming into your home from the electrical utility). Second, generating solar energy from panels is a zero-emission source of electrical power.

Having solar panels on the home is metaphoric to having wood to chop. Both bring energy supply, and the knowledge of how it works, closer to home.

ⓠ FACILITATOR'S TIP

Have the group discuss this "intimacy" with an energy source. How close do they feel to the energy sources they rely on?

---

1 E **What are some challenges in educating the public on energy supply?**

I've followed this issue of "energy literacy" for many years. Few people question the need for a greater understanding of how our energy works. However, the reality is that, for most people, how the lights go on or the furnace fires up is not top of mind. What's important is that a flick of a switch turns them on.

Yet, the sourcing and delivery of energy to power our amenities encroaches on every corner of society. For example, electrical lines, solar panels, wind farms and oil wells clutter the landscape, and pipelines carry fuels under our feet. But as you taxi out to the runway for takeoff, do you think about where the plane's fuel came from? Probably not, and you're not alone.

When the discussion turns to who should be responsible for teaching the subject, fingers often point to energy suppliers such as oil and gas producers, pipeline companies or electrical utilities.

I tend to agree that energy suppliers, like any business, must play a role in educating their customers. Yet corporate efforts have been poor or, in the worst case, completely mistrusted by the public. In not getting its message out earlier, the industry is now scrambling to communicate how it's addressing key issues, such as environmental concerns. In the meantime, its detractors have been executing a concerted campaign for years. It's tough to come out from behind, and energy producers have a long way to go before their communications aren't perceived as pure propaganda.

Educating the public about energy is difficult. Generally, people want to know their energy is "cheapest, safest, cleanest and quickest" without having to understand what goes on behind the scenes to deliver that. Yet, to achieve sustainability, it's incumbent on us to know more, so we can be more responsible in our choices. Teaching such personal responsibility is probably best started at an early age, in homes and schools.

To see how important early education can be in forming a society's energy worldview, check out another Energyphile story, "Once Upon a Time ..."

⚗ FACILITATOR'S TIP

Have the group discuss the government's responsibility in educating the public on its energy supply.

---

2 A Why do consumers today receive subsidies to install solar panels on their homes or buy electric vehicles (EVs)?

Government intervention with policy tools like subsidies is necessary if the intent is to accelerate the adoption of new technologies.

Subsidies lower the price of solar panels and electric vehicles, thus stimulating faster adoption of energy systems that either were, or still are, uncompetitive against the incumbent. For example, EVs are largely too expensive relative to a combustion engine car, so it takes a subsidy stimulus on the new (in conjunction with a carbon tax on the old) to economically induce more people to buy the new.

The more people buy solar panels and EVs, the more their suppliers can scale up operations and reduce costs. After a certain point, the cost of these new products should fall below their incumbent competitors, and, in theory, no more stimulus should be required.

So, subsidies are meant to work both for the short and long term. Short term, they push adoption enough to create free-market momentum that enables companies to establish large-scale capacity. This capacity in turn feeds mass-market adoption in the long term. That's how the electric car and its associated roadside charging infrastructure are playing out.

2 B Why didn't Alfred get a subsidy?

A hundred years ago there wasn't an environmental impetus to accelerate a transition from wood to gas. Hence, the adoption rate of gas appliances and infrastructure was left up to the free market. People switched from wood to gas progressively faster as the price of gas fell to a level that was compelling enough to send customers to appliance stores like the one on Gas Lane.

The difference between the energy transitions we desire today and many that happened in the past is that, now, environmental urgency drives many of them. Consequently, subsidies are implemented to usher these transitions along.

2 C Do you think a wood tax would have influenced Alfred to switch to gas faster?

In my opinion, probably not. Alfred was wealthy, and the wealthy are typically insensitive to small changes in commodity prices. If a wood tax were to have been imposed, it likely would have been at an amount inconsequential to Alfred and his ilk. If the tax had been too high, it would have disenfranchised lower income people, who couldn't afford higher wood costs, never mind expensive new appliances and the price of gas.

3 A What's a current example of wealthy people being the first to adopt new green energy–consuming products even if they're more expensive?

The best example I can think of over the past decade is Tesla's electric vehicles. The company's first sedan, introduced in 2007, was the Model S — a US$100,000+ luxury vehicle only within reach of high-income buyers. At the time, it was also an unknown quantity relative to its equivalent luxury peers offered by the likes of Cadillac, BMW, Audi and Lexus.

As more people bought the Model S, followed by the Model X luxury SUV, Tesla's operations became more streamlined. Scaled manufacturing processes dropped costs, and adoption began to accelerate. By the time the more affordable Model 3 was introduced in mid-2017, adoption of electric vehicles had taken root and moved into the mainstream — no longer were they accessible to the rich only.

3 B What about organizations? First, think about examples around the world, then discuss what your organization is doing, if anything, to adopt new energy technologies.

Notable examples in the 2010s include cash-flush companies Microsoft, Apple and Google. Recognizing their insatiable need for electric power — to drive server farms and other energy-intense computational processes — they and other companies began building out directly accessible renewable energy projects to bring down their reliance on fossil fuels.

3 C At the global level, what countries have been leaders? To what benefit?

In response to calls to address climate change in the early 2000s, Germany, Norway, Denmark, the United Kingdom and other European countries began taking leadership positions in adopting renewable energy technologies. Much of the early buildout of wind and solar projects, between 2005 and 2015, was not economically viable and needed substantial subsidization. China followed suit in the 2010s, as did other countries, including the United States and Canada.

These countries' early investment and adoption led to rapidly falling costs for manufacturers and installers, who were able to scale up operations and improve technologies. By the latter half of the 2010s, solar and wind power became operationally competitive with incumbent power-generation systems like coal and natural gas. Similar trends are being shown for energy storage, for example large-scale batteries.

3 D Discuss the role of income inequality in facilitating energy transitions.

Offering new products to society's wealthy segments first has proven to be a successful business strategy. Those with pools of money provide the most fertile ground for early adoption — it's common sense to target them.

Following the money is not a new concept. In the 1880s, Thomas Edison offered his novel light bulb and electrical systems to Wall Street first. Again, adoption by society's wealthy segments helped induce scalable operations and drive down costs. After a certain point, costs fall enough to be competitive against the incumbent. And if the utility is there and it's a better product offering, adoption will permeate into broader society.

In this regard, income inequality is leveraged from the top down to bring new products to market. When those products are for the betterment of society (like reducing environmental degradation or bringing people out of poverty), it's to everyone's benefit. Recognizing this dynamic is important because income inequality can act as a free market mechanism for wealthier segments of society to fund the early learning curve effects (cost reduction) for benevolent product introductions.

This thinking applies at a national and global level too. Note that in question 3c, all the countries are wealthy. When the world is trying to accelerate an energy transition, understanding the dynamics of funding early product and process introductions is vital.

💡 **FACILITATOR'S TIP**

Have the group discuss the role of income inequality in *resisting* energy transitions.

4 A Discuss the reasons for the deflationary trend in energy pricing.

Energy has become cheaper over time for three primary reasons: improved production technologies, a shift in the source of gas and the scaling up of infrastructure and operations.

As discussed earlier, scale drives down costs. A hundred years ago, pipelines and gas delivery were new in Nova Scotia, so the first customers had to pay more. As more customers came aboard, and gas production and distribution capacity were built out, the costs for those who supplied the energy came down. As marginal costs came down, so did the price to consumers.

The use of gas in North America didn't really take off until post–World War II. The 1950s were the glory days of adoption. Between the early 1900s and mid-century, the source of the gas delivered to homes, dominantly methane, transitioned. That behind-the-scenes change — invisible to gas consumers — helped drive down costs (hence price) too.

The type of gas delivered to Alfred Dickie's home was coal gas, or "manufactured gas," the latter ascribed because the gas was manufactured (gasified) from coal. Basically, lumps of coal were heated in a closed oven. This foul process liberated combustible gases, mostly methane. The gas was "washed" and put into the pipeline for distribution to customers.

In the 1930s, a new form of cleaner gas, also mostly methane, was brought to North America through long cross-continental pipelines from distant oil and gas fields. This new gas, called "nature's gas" or natural gas (because it came from nature's underground reservoirs), displaced the more expensive, dirty coal gas. Over the course of the 20th century, this upstream transition from manufacturing gas in coal ovens to drilling natural gas from wells led to greater scale, lower costs and therefore progressively lower consumer gas prices.

4 B **Do you expect the real price of energy to continue its decline over the next hundred years? Why or why not?**

The trends suggest the answer is yes, although it's hard to get much lower, because in some places in North America, natural gas comes out of the ground for almost free.

In the early 21st century, another behind-the-scenes process transition, called shale gas, began displacing natural gas extracted from vertically drilled wells. Shale gas is natural gas trapped in extremely compressed rocks called, not surprisingly, shale. Starting around 2005, the extraction methodology changed dramatically, enabling producers to drill down to the shale layers, then out horizontally a few kilometers into the layer to get maximum exposure to gas reserves. Through a process called hydraulic fracturing, the shale layer is shattered and the highly pressurized natural gas is liberated in great quantity.

This new drilling and completion process has resulted in so much new gas production capacity that regional prices in North America have plummeted to as low as $1 per thousand cubic feet, or less, at the wellhead. Without correcting for inflation, that's the same price Alfred Dickie paid a hundred years ago!

That's natural gas. Post-2010, the price of renewable energy —
especially wind turbines and solar panels — began to fall dramatically. In less than a decade, technological advances in both
drove costs down to achieve a long-desired goal: parity with
natural gas power generation. Over the next decade, renewable
energy technologies will continue to improve in efficiency and
cost, and their utility enhanced with power storage.

We are entering a period of abundance, where the ability to
bring a joule of energy to consumers — from all primary sources —
is becoming easier and less costly. For consumers, this means
cheaper energy costs over time (in real-dollar terms).

4 c **Discuss the economic and political factors that can limit
forever declining energy prices.**

Economically speaking, the theoretical limit for how low the
price of energy can go is the marginal cost of its system. For
natural gas, that's the sum of all costs for bringing to market
the last cubic foot demanded by customers. At some point, the
extraction and distribution technologies of the day, combined
with other supplier expenses, including labor, maintenance and
taxes, cumulatively define a floor for the marginal cost of any
energy source.

Politically, a jurisdiction's resources are owned and governed
by the state. In other words, on behalf of its citizens, the government is the custodian of a resource like natural gas. From
this perspective, it's not in the state's best interest to sell its
resources at progressively lower prices. That's why many countries, states and provinces have a severance tax, which is a tax
on production (or severing the resource from its reservoir).

Royalties are a share of the profits that go to the resource owner, which, from a producer's perspective, show up as a cost.

For renewable energy systems like solar and wind power, in addition to the cost of operating the panels and turbines, respectively, there are costs associated with the acreage they occupy. Payments to landowners and taxes to local governments constitute fixed costs that contribute to a floor price the supplier must charge to make money.

All energy supply systems also necessarily come with physical equipment and machinery. The cost of maintenance is not trivial, especially in harsh climates.

Taxes typically constitute a large fraction of costs for any energy supplier. And for fossil fuel suppliers, the specter of increasing carbon taxes is a countervailing force to otherwise falling costs achieved through technology. Carbon taxes are meant to elevate the cost of fossil fuels, to more than offset the tendency for technology to pull down costs, and ultimately keep prices high enough to encourage customers to switch to renewable energy alternatives.

4 D What are the negative consequences of continually declining consumer energy prices?

Continually declining prices mean consumers will buy more primary energy and be less inclined to responsibly consume the work that's derived, for example, heat, light and mobility. The cheaper an energy source becomes, the less incentive there is to be mindful about conservation and efficiency. This principle applies whether we're talking about fossil fuels or renewable energy.

For example, continually declining solar panel prices make this non-emitting energy source more competitive, hence more adoptable, which is a good thing. But it's a double-edged sword. With more consumers of solar power, and with its price on the decline, indiscriminate consumption results, necessitating the manufacture and installation of more panels, which are made from non-renewable extractive industries (mining). And covering acres of arable land with panels simply to satisfy irresponsible consumption habits is not in the spirit of sustainable solutions.

The bottom line is that there are environmental and social costs to every primary energy source. Allowing energy to be too cheap leads to unsustainable outcomes.

5 A **What attributes are most important to customers buying appliances today?**

I think it's obvious to say that cheapest, safest, cleanest and quickest are still top dimensions of utility.

We can add "most reliable" to the list. Today, the expectation that the lights, stove, furnace and other appliances will turn on when the switch is flipped is deeply entrenched.

In addition, "greatest efficiency" is a key factor. Historically, efficiency — the amount of energy consumed to do the work needed — would have meant being cheap to operate, because greater efficiency means lower operating cost. However, efficiency today also encompasses the environmental dimension, because lower energy use implies lower greenhouse gas emissions too.

5 B Do we take for granted the amenities that energy brings to our lives?

I believe that many of us in developed, wealthy countries do take our energy for granted and expect it, and the appliances it powers, to be cheap, safe, clean and quick.

An optometrist's cliché is that the best eyes are the eyes you don't have to think about. It's the same with energy — you shouldn't have to think about it. You flick the switch and your hair dryer starts. You arrive at home and it's the perfect temperature. You don't worry about being electrocuted when you plug in your phone, or choking to death when you light the stove. These are all attributes we expect. In fact, we expect them to the point we're complacent about them.

On one hand, that's a good thing, because it's an indication of how reliable our energy systems are. On the other, it makes us lazy in understanding issues that may be hidden from us, or that one day may surprise us. Often, we only appreciate the benefits of something when it's taken away.

For more on how hidden issues, combined with our own self-interest, can affect our energy judgment, see the Energyphile story "Nobody Tips a Scandiscope."

💡 **FACILITATOR'S TIP**

Get the group to discuss the last time they experienced a power outage. How did it feel? How about the last time they experienced a price hike on their utility bill?

---

5 c Are you willing to trade off "cheapest, safest and quickest" for being a zero-carbon emitter, or "cleanest" in today's terms?

Trading away safety and convenience ("quickest") is pretty much non-negotiable for most people. As discussed in question 5b, our expectations regarding the safety and convenience of our energy systems are deeply, deeply entrenched. If people are to adopt a zero-carbon alternative, it had better be just as reliable as their current system.

But what about price? In the pursuit of cleaner energy, is "cheapest" negotiable?

Some countries, most of them in Europe, have embraced higher energy costs as a means to encouraging conservation, efficiency and switching of fossil-fuel energy sources in the pursuit of curbing emissions.

But accepting higher energy costs to effect a transition to cleaner systems is not universally palatable. For instance, the implementation of a carbon tax has incited protests, even a change of government, in some jurisdictions.

Before meaningful emissions reduction can occur, there must exist a culture of wanting to trade "cheap" for "clean" at an individual level.

Renewable energy systems are innovating quickly, so making the trade-off to clean without paying more, or without sacrificing safe and quick, is becoming easier. However, I ask this question because it still costs money to shut down, for example, a functioning coal plant and build wind turbines. Taxes to force switching also cost people money. As discussed in question 3, wealthy countries and individuals are able to do this, but such access to capital is not universal. Those without money still must grapple with the most contentious trade-off: cost versus clean.

---

6 A If you were an investor in 1915, what considerations would you take into account before investing in the gas business?

For any new consumer product (in this case, a gas stove) and process (burning gas instead of wood), the first point of investor skepticism is whether people are willing to change their entrenched way of doing things and switch to the new.

The propensity for a consumer to switch depends on many factors, including the relative *utility* of the new versus the old. And, of course, the price difference between the benefits — whether the improved appliance is worth the price on every useful dimension. As an investor, you can't be analytical enough when assessing the relative utility of new versus old.

The key dimensions of utility the Halifax Electric Tramway Company offered are spelled out on the bill: cheapest, safest, cleanest and quickest. Was all that really true relative to burning wood in a stove? Was the quantum of difference in things like safety and cleanliness compelling enough for a household to pay up for a new appliance and monthly service? If, as a potential investor, you felt that using gas was merely as good as, or only slightly better than, burning wood, you might have decided the incentive to switch was not there. And, therefore, neither was your investment opportunity.

As an investor in 1915, you should also have asked yourself if there were other dimensions of utility. A key consideration is not listed on the bill: reliability. Or, at least, perceived reliability. Back then, people took comfort in looking out the window and seeing how much wood they had stored for winter. So, you should have been skeptical about whether homeowners would be willing to give up control of their primary energy source. I imagine one of the first things a potential customer might have asked was whether the Halifax Electric Tramway Company would be able to deliver heat reliably — as reliable as a stack of wood in the backyard — on a cold winter day.

Then there's price, which is a measure of what consumers are willing to pay for the utility of what's offered. As a potential investor, you'd have to pay attention to whether gas stoves and piped-in service were priced low enough to lure consumers to the incremental benefits being offered over wood.

Once convinced of the venture's commercial potential, you'd need to consider whether the company could expand to meet growth and other operational considerations.

Of course, assuming all else checks out, the ultimate consideration is whether the company can make money and deliver returns to shareholders.

We know how history played out: in all respects, gas stoves had compellingly better utility over wood. The Halifax Electric Tramway Company lived up to its claims, justified its prices and made money.

6 B What was it about gas stoves that made them so successful over wood?

I think "quickest" was one of the most compelling features that drove consumer adoption. Getting instant heat by turning on a valve and flicking a match, and then turning off the heat equally quickly, was a huge selling feature that customers were prepared to pay a premium for. In addition, there was time savings in not having to go out and chop wood, especially in cold weather.

Being "cleanest" was an important factor too. Not having to scoop out ashes would have been a pretty big selling feature, though probably not as compelling as the "quickest" dimension.

Better safety was questionable. Potentially leaking gas pipes and the associated risk of explosion were not demonstrably safer than a fire carelessly started with neglected embers.

Over time, the price of stoves and gas declined. That, combined with the net benefits of using gas over wood, made for a successful transition.

6 C What are some new appliances or products that didn't succeed on the market?

There are many, many examples over the past hundred years. More recently, there's Google's Glass, LaserDisc, Apple Newton and 3D television.

The one product that's been the most interesting to watch is the electric car. There have been several attempts to displace the market dominance of petroleum-powered vehicles, notably in the 1970s and 1990s, but each ended in failure. Only recently have electric cars demonstrated convincing market penetration and potential. The conditions are looking better than ever for widespread adoption.

Note that while the Apple Newton may have failed, it was a harbinger for the iPad. And the probability we'll be wearing computerized eyewear in the near future seems high.

So, although a product doesn't make it the first time (typically because it doesn't have a compelling set of benefits over the status quo), it can eventually gain the qualities needed to woo previously skeptical customers, and investors too.

6 D What have you learned about new product adoption?

I wrote "Alfred Dickie's Utility Bill" in part to help you think about what prompts consumers to make the switch from one product and system to another. The factors that facilitated the switch from wood to gas can help us understand to what extent citizens today may make a switch to new modalities, for example, electric cars.

The primary lesson is that people will make the switch to a new product and system if the net benefits of the new product *compellingly* outweigh the net benefits of the old, and if the price paid for that difference in benefits is perceived to be worth the change. As price comes down, the attraction to switching goes up. This is a fundamental principle in assessing the viability of products and processes in today's desire for energy transition.

7 A Why do you think the Halifax Electric Tramway Company offered such a hefty discount for early payment?

World War I had started a year prior and interest rates were going up. The value of money — hence the cost of servicing corporate debt — was increasing. I don't know how much debt the Halifax Electric Tramway Company had taken on to build out its gas piping and electrical wire infrastructure, but I suspect it was a fair bit. A debt burden combined with rising interest rates probably meant that getting cash in the bank was important. Offering a discount to customers so they would pay as soon as possible was a prudent tactic.

But being able to give a steep discount also likely meant the company was profitable and could afford to sweeten customers' bills. It's usual for first-to-market companies, especially in an industry that's difficult to break into, like a utility, to enjoy higher profitability in the early stages of commercialization (with the proviso that their offering has compelling enough benefits to attract buyers, as discussed in previous questions).

Over time, new suppliers come into the market, competition gets stiffer, and prices fall. I would expect that a utility bill from the 1920s would not offer customers such a generous incentive to pay early.

7 B What are the external influences that companies offering new energy solutions today must contend with?

As we saw in the last question, interest rates, availability of capital and war are some of the many macroeconomic factors that can affect the adoption of new products.

Inflation and unemployment rates are factors too. All else being equal, the less healthy the economy — with characteristics like high inflation and unemployment — the more difficult it is to commercialize new offerings.

The competitive landscape is another important player in the early days of adoption. If there is little to no competition to start, the companies with hot new products can charge more at the outset.

I find that people who talk about energy transitions often focus too narrowly on technology and how new product introductions are a "no-brainer" for consumer adoption. But even if the overall net benefit of a new product over the status quo can be demonstrated, I always advise companies that macroeconomic conditions must be considered. And the level of competition dictates pricing strategy, including the potential to offer customer discounts.

Today, interest rates are extremely low and the global economy hasn't seen a recession in over a decade. That means money is relatively plentiful. For now, such conditions are conducive for building large-scale renewable energy projects — that's because 75% or more of such projects are financed with debt. But what would happen to the propensity to spend (corporate and personal) on new things if interest rates were to escalate or money became constrained due to recessionary forces? Where would a government's fiscal policy priorities lie if unemployment were to rise?

Always assess macroeconomic factors, especially the cost and availability of capital, when thinking about the adaptability of new energy systems.

💡 FACILITATOR'S TIP

Ask group members to discuss what macroeconomic forces affect their own organizations.

FACILITATOR'S GUIDE

Come Together, Move Forward

Whether it's a corporate planning session, a class discussion or a social book club, an Energyphile Session encourages critical thinking, sparks lively conversations and helps build a community equipped with tools needed to make thoughtful, well-informed decisions for a better energy future.

The questions in this discussion guide invite participants to dig deeper into the issues explored in the story and gain a broader understanding of the forces of change that affect our energy circumstance. Encouraging people to learn from the past, understand the present, then prepare for the future, these questions embody the Energyphile philosophy.

As the facilitator, you will ensure the conversation stays on track, its objectives are met and everybody walks away with a better grip on how they can move forward, while having had stimulating conversation — and some fun — along the way. This guide will help you plan and prepare for a fruitful discussion.

Planning

WHAT'S YOUR OBJECTIVE?

Start by defining your goals. Every gathering of people is different. Will this workshop be used as part of a team-building program? Is your organization facing a particular business issue you'd like your C-suite to tackle in a strategy session? Or do you just want to provoke thought at a family barbecue? Whatever the reason, bringing people together fosters a greater sense of mutual understanding.

WHO SHOULD YOU INVITE?

The occasion and objectives will determine who to include. If your guest list is less prescribed, consider inviting people from different backgrounds, companies or departments. One of the more interesting discussions we've seen included employees from across a company, both office and field staff. Another robust session saw a group of accomplished friends gather around a dinner table.

HOW MANY PEOPLE?

Aim for 6 to 10 people, preferably not more than a dozen. Any more and you can separate into breakout groups. Smaller groups allow for more questions to be covered with greater depth.

HOW LONG?

A good length is 2 to 3 hours. If you want to offer a longer workshop, consider breaking it up into discrete sections — for instance, cover all policy-oriented questions in one — or incorporating other activities, like having groups research different questions and report back.

WHERE?

Of course, there's always the boardroom, but think about leaving the office. Getting people out of their usual environment and routine turns it into a more social outing and may help shift the dynamics. Many coffee shops and restaurants have private rooms. The library probably rents rooms for free, or there may be innovation hubs that offer space. Check out community centers or coworking spaces for rentals, too.

If you're venturing outside the office, ensure the space can accommodate your tech requirements and other amenities, such as catering. If you're rounding up a geographically scattered group, try Skype, Zoom or another video-conferencing app.

Preparing

Use this checklist of tasks and suggested timeline to prepare for the day of discussion.

TASK	LEAD TIME	COMPLETED
Book the venue and any required tech.	2 months	
Invite participants (request RSVP).	6 weeks	
Order hard copies of the discussion guide for participants from Amazon or Indigo.* (If participants are responsible for purchasing their own, ensure they've ordered at least two weeks in advance.)	6 weeks to 1 month	
If you plan to record the workshop, take notes or generate action items, delegate someone to be responsible for that.	1 month	
Order catering.	3 weeks	
Distribute discussion guide to participants.	2 weeks	
Email the energyphile.org link to the story so people can read its vignettes and listen to the audio version. (Some will prefer audio over print.)	2 weeks	

* For orders of 10+ copies, Energyphile offers a discount. Your order must be received at least six weeks before your event. Contact hello@energyphile.org.

TASK	LEAD TIME	COMPLETED
Email any other supplementary material participants should read in advance (news stories, internal documents).	2 weeks	
Read the introduction to the questions to understand their general considerations and themes.	1 week	
Read the questions, prioritize them based on your objectives and allotted time. Add any of your own.	1 week	
Organize the tech and tools you'll need (whiteboard, flip chart, markers, sticky notes, laptop, projector, screen).	5 days	
Finalize other material you intend to use (PowerPoint deck, handouts).	4 days	
Send a reminder to participants to read the story and questions (but not the answers!). If you plan to cover select questions, you may want to let participants know which to focus on. Remind them to bring paper/pen or laptop for note-taking if they desire, and of day, time and venue. If you plan to record the session, note that, too, in case anybody has concerns.	3 days	
Refresh yourself with the questions you've identified as your priorities and determine how long to spend on each. Think about how you'd like the discussion to unfold.	1 day	
Make name tags or cards for all participants if they don't know one another.	1 day	

During

Kick things off by stating what you want participants to gain from the discussion: Learn takeaways they can apply to their day-to-day role? Determine a course of action for a specific strategic initiative? Or is it purely meant to get brains working and people talking? Ask the group what they hope to get out of it, too.

Cover the housekeeping considerations: how long the discussion will last, format, when breaks will happen, where washrooms are. If you intend to record, remind people of that.

Ask everybody to briefly introduce themselves: name, where they work/what they do, what they hope to learn.

Establish the ground rules:

- Respect all points of view — maintain an open, supportive forum for every person to express their thoughts and to learn from one another.

- One speaker at a time. Give people space to speak.

- Avoid side conversations or other interruptions.

- There is no right or wrong. Disagreement is welcome but don't make it personal. (Address the *idea*, not the person who shared it.)

- Mute your phones.

As you guide the group, weave the past, present and future throughout the conversation. Remember the Energyphile philosophy: learn from the past, understand the present, prepare for the future.

If action items will result from the discussion, ensure these are documented as you go.

As the discussion progresses, watch the time to ensure you're sticking to the pace you've determined.

Allow about 15 minutes at the end to review the action plan and summarize the discussion's main takeaways with the group. What were their most surprising "Ahas"? Did their views change as a result of the discussion?

TIPS FOR MODERATING

Your role is to keep the discussion on track while creating an environment that ensures all participants feel comfortable to speak their thoughts, even if dissenting. These tips can help you do that.

People may need time to warm up. If that first question has you facing a silent room, try paraphrasing it or suggesting other ways they may come at it. If you know someone has expertise or interest in the area, invite them to share their thoughts if they're comfortable. You could also move on to another question — perhaps one that relies more on subjective takes than deep subject knowledge — and return to the first later.

Guide the conversation to keep participants engaged and focused (very important!). People can easily get sidetracked when exploring contentious questions — make sure they stay on topic.

If the discussion does become difficult or tense, pull it back to the story to re-establish common ground and remind people of the lesson.

Direct the conversation, but don't dominate it. Stay neutral and focused on listening, rather than offering your own opinion.

Make sure everyone has a chance to be heard — and understands the gift of listening. This may mean you have to clear space for people. Moderate frequent contributors if they start to dominate, and stay attuned to the quiet participants. Watch body language. Does somebody seem uncomfortable with a topic or speaker? Does a normally silent person look like they have something to say? Use those cues to steer the conversation.

Encourage people to unpack their responses when appropriate. Prompt them with questions like "How?" "What leads you to that view?" "Can you give me an example?"

Use natural segues or lulls in conversation to move on to the next question. If the discussion shows no signs of winding down, blame the clock for cutting it off and proceeding to the next question.

After

Within a week, follow up with notes, the recording, action items and anything else promised.

If you're considering offering such sessions regularly, send participants a survey to gauge what worked, what didn't, areas for improvement or change and preferred next topics or stories.

Finally

Enjoy the discussion!

SOURCES CITED
AND
IMAGE CREDITS

Sources Cited

Page 21: "The remaining 18% of Canada's electricity ..."
Statistics Canada, *Electric Power, Annual Generation by Class of Producer* (Table 25-10-0020-01) 2017
j.mp/statscanadaelectric

Page 22: Electricity Generation by Fuel, 2018
BP, *BP Statistical Review of World Energy 2019*
j.mp/bp2019review

Image Credits

All objects shown are in the collection of Peter Tertzakian. Photography by Peter Tertzakian: page 13. Other images scanned from original sources.

ABOUT THE AUTHOR

THE QUINTESSENTIAL ENERGYPHILE, Peter Tertzakian has devoted his career to energy, first as a geophysicist, then as an economist and investment executive. He's written two bestsellers — *A Thousand Barrels a Second* and *The End of Energy Obesity* — and is sought around the world as a trusted, engaging speaker. Energyphile is the culmination of his passion and knowledge.

DISCOVER MORE AT
ENERGYPHILE.ORG

- Let your curiosity wander in the PhileSpace museum
- Explore artifacts in the Energyphile collection
- Hear the stories come to life in the audio productions
- Check out new stories and content

Continue the conversation with Energyphile Sessions

If you enjoyed this book, explore others in the Energyphile Sessions series. Covering a range of issues, these discussion guides will get you thinking about the business of energy in a whole new way. More are always being added, so check back regularly.

energyphile.org/sessions

Manufactured by Amazon.ca
Bolton, ON

12569430R00039